SPACE MEDICINE

SPACE MEDICINE

Written by:
Austin Mardon, Catherine Mardon, Maya Nagorski,
Amna Zia, Monica Thakadu, Kirithika Bharatselvan,
Brianna Bedran, & Botho Modutlwe

Edited by:
Anastasiya Yermolenko

Copyright © 2021 by Austin Mardon
All rights reserved. This book or any portion thereof may not be reproduced or used in any manner whatsoever without the express written permission of the publisher except for the use of brief quotations in a book review or scholarly journal.
First Printing: 2021

Typeset and Cover Design by Kim Huynh

ISBN: 978-1-77369-639-3
EBook ISBN: 978-1-77369-640-9

Golden Meteorite Press
103 11919 82 St NW
Edmonton, AB T5B 2W3
www.goldenmeteoritepress.com

Table of Contents

1. What is Space Medicine? 1
2. Human physiology on Earth vs. in space 7
3. Mental health and the human mind in space 13
4. History of advances in space medicine 20
5. Pathophysiology in Space 26
6. What are the implications of space medicine on Earth? ... 31
7. Medical specialties in space 37

Chapter 1

What is space medicine?

Zain Kadri

Space medicine is the practice of medicine in outer space. Space medicine is crucial for all space missions, prior to launch and after launch. They are especially used to diagnose, or treat any underlying conditions an astronaut may have prior to launch, which may possibly affect the astronaut on his mission.

It is important for an astronaut to be healthy on all space missions before launch as he or she may be negatively impacted throughout the entirety of the mission. In addition, a surgical team, with health expertise must be present during the trip at all times to ensure the safety of crew members (NASA). Also known as flight surgeons, these crew members are specifically brought on to missions to provide urgent care, this can range from anywhere from a minor cut to deep gashes and near death occurrences. Specifically, a flight surgeon is in charge of the following: 1. Ability to diagnose, treat and prevent cardiovascular diseases, 2. Ability to understand, and discover new approaches to treat osteoporosis, 3. Early detection of birth defects, 4. Emergency medical care to any diagnosis and 5. Treatment of metabolic disorders inside and outside the aircraft (NASA).

Space medicine, specifically engineered for astronauts going to space, has its advantages and disadvantages. Before astronauts go to space, they are required to take a drug to increase blood pressure temporarily so they do not get dizzy. This drug is called midodrine ($C12H18N2O4$), approved by the United States Food and Drug association in 1996 for dysautonomia and orthostatic hypotension (Mitarai, G). This medication must be taken with caution, and in a controlled environment, usually prescribed to be taken 3 times a day, with or without food. The special drug is not limited to only astronauts; many people are prescribed this drug to prevent low blood pressure and other underlying conditions (Pass, J). This drug can be harmful to some people with the following conditions: severe heart disease, having an adrenal gland tumor, kidney diseases, inability to urinate and having high blood pressure while lying down.

Space medicine also consists of being physically fit as well as maintaining proper standardized hygiene. Astronauts must be able to do cardio and be able to have function in all joints (excluding disabilities). Physically, astronauts should be fit according to a certain standard. Astronauts are strictly required to have good health and follow the given criteria. The touchstone of your height must be 4'9, and the max should be 6'2. Your weight must be within the limit of 210lbs and not less than 110lbs (Matsumo). Astronauts are obligated to have 20/20 vision (with glasses) and should not be colour blind. Future astronauts should be able to hear perfectly, with no chronic or recurring auditory circumstances. Finally, astronauts must not have blood pressure over 140/90mm in a sitting position. (CSA)

All candidates must pass the European part MED class 2 examination by a professional and certified medical examiner (CSA). All applicants with disabilities must have a physician to be able to give a medical certificate stating that without their underlying condition, they are able to abide the European Part MED class 2 examination (NASA). Applicants must be free of any disease, addiction or dependency on drugs, tobacco or alcohol. Fluidity and joint movement should be normal (specific disabilities are exceptions), and no psychiatric disorders should be present either. (NASA)

Why is space medicine important?

Space medicine is fundamental to human flight to space. This helps astronauts overcome internal challenges to adapt into a complete new environment in outer space. Not only does it boost cognitive and physical performance, but it enhances your chances of survival and contributes to your basic functions (Nasa). The main reason space medicine is required before flight is because in outer space, blood pressure is lower compared to here on Earth (The Conversation). To help prevent low blood pressure, which can lead to passing out or other minor symptoms, it is vital to take space medicine in a respected amount and in controlled doses.

History:

In 1948, a former Nazi by the name of Hubertus Strunghold constructed the term Space Medicine as he was the first and only professor in Space medicine in the School of Aviation Medicine at Randolph air force base, Texas (NASA). While constructing new suits, he was asked to create medicine to further boost astronauts' performance and ensure mental health would be in check before flight/ take off. Hubertus Strughold contributed the entirety of his life to aviation sciences due to the love he had for physics, science and medicine (Lipinski, C).

Medical Equipment aboard a spaceship:

The required necessity of a spacecraft is having emergency medical equipment on board in case of emergency situations. On a spacecraft, IRED, a device specially equipped to allow astronauts to work out in space (NASA). Astronauts work out a minimum of 2 hours a day with IRED, to ensure their bodies do not lose mass. A spaceship comes with a med kit, to treat smaller issues, however, surgical equipment is also brought on board to treat deeper wounds/ gashes. Astronauts will usually feel symptoms like nausea, motion sickness, back pain, burns and dental emergencies. Astronauts are also trained on how to correctly surgically extract a tooth on a functional human being before take off (Lipinski, C). The medical kit on board a spacecraft generally contains basic medical tools such as a defibrillator, portable ultrasound, a device to look deep within the eye and 2 liters of saline. A book containing general conditions is provided to gather and piece information when a fellow astronaut is feeling sick or has a wound. The book provided should establish an understanding to the reader with what he or she may be experiencing in relation to their symptoms.

Medical symptoms may vary in space, and it is important that reliable research be done prior to launch to ensure astronauts are given the right information to deal with problematic issues effectively. NASA, the space research centre founded by Bill Nelson, gained funding from the United States government, which dedicated over 600 billion dollars to research alone. Further research is needed to succeed in future missions and soon be able to inhibit other planets. By 2050, humans should be gathering more knowledge to soon be able to land on Mars and other inhabitable planets which we are currently in the process of understanding and researching.

When there is no gravity, your muscles feel a lot looser, as they are not used as much. Compared to Earth, where we are always pushed down because of gravity (American Psychological Association). In space, the importance of working out is highly prioritized in order to be able to have full functioning body parts and fluidity in joints. You may be wondering how these astronauts work out without the force of gravity. The simple answer is: They create it artificially. To create artificial gravity, you must use some sort of centripetal force. It is a mechanism which circles around your spacecraft, which adds a sort of weight to the astronauts inside the space station. By adjusting the parameters, radius, rotation rate and adding perimeter, you can adjust the weight similar to gravity (NASA). Before you know it, the spacecraft will feel like home- neglecting the fact that you are 528km in the sky (NASA).

The next problem astronauts often encounter is the fact that radiation is constantly being transmitted throughout the spacecraft (American Psychological Association). Radiation in space is worse in space compared to earth, so obviously they had to find a way to limit the exposure to constant radiation. When astronauts go on long trips they are affected negatively to the constant amount of radiation transmitted while on board. To limit exposure, a sort of shielding is layered over the spacecraft using hydrogen that can absorb cosmic rays (American Psychological Association). Pure Hydrogen cannot be used to layer the spacecraft, so instead they use materials high in hydrogen like polyethylene or a common plastic containing one carbon and two hydrogens. This material is heavy and only blocks 30% of radiation transmitted. The remaining 70%, astronauts must deal with themselves using antioxidants like vitamin A and vitamin C. NASA scientists are currently in the process of researching ways to help the body after damage has already been done. They are looking for ways to make cell repair easy within the body in the most natural way possible. They try to give the cell more of a chance to repair its own problems.

Future medicine, NASA aims to have robots perform surgery in space. This would limit the amount of training and burdens astronauts would have to go through, as human error is possible. Vice versa, the machinery must be in perfect use before being put to the test as tech, coding and calibration error is also able to be made (Nasa). Robotic precision always beats the precision of a human hand. Humans make more errors than robots, so it is more likely that they will replace human surgery with robot surgery. Human surgery will still be needed for smaller decisions and cuts. The use of tech is changing the world as we speak, and will most likely replace multiple jobs at once. As the years go by, more and more people will be unemployed and will have to adapt to technology. The future skill that most individuals will have to pick up is adaptability. Due to the rapid pace and change, we must be able to adapt and overcome setbacks like these. The same goes with astronauts as some of them might lose their jobs (such as surgeons) because robots will be able to do the same job for half the price and half the time.

The importance of studying space medicine seems quite clear as of now. We are currently on the brink of a great breakthrough in the space medicine field to further conduct research and make our lives easier. We must continue and strive to help research the topic of space medicine, as it is still underdeveloped.

Chapter 2

Human physiology on Earth vs. in space

Maya Nagorski

Introduction

Space travel is a dangerous pursuit; perhaps for no reason so much as the devastating effect it can have on human physiology. The conditions on a spacecraft are very different from those on Earth, and can put enormous strain on one's body, as proven by numerous publications from NASA and various respected medical professionals (Williams et al., 2009). Factors such as microgravity, the presence of a vacuum, extreme temperatures, and cosmic radiation can lead to serious consequences such as muscle atrophy, loss of bone density, failing immune systems, and more (Williams et al., 2009). These side effects can influence the rest of an astronaut's life, long after they have finished their voyages in space. There are treatments to remedy some of these conditions, though there is always a need for more research and more solutions. With space travel becoming more and more ambitious, and the daunting possibility of long-term flights to other planets within the solar system, this is becoming more pressing than ever. This chapter will focus on the dangerous conditions astronauts face, the effect this has on their bodies both in the short term and long term, and finally, the steps required to recover once back on Earth.

Microgravity

Microgravity is defined as the feeling of weightlessness experienced when there is minimal force of gravity (Dunbar, 2012). The results of this phenomenon include exceptional disturbances to one's blood plasma, internal fluids, muscles, and kidneys (Pal et al., 2021). This, most notably, affects bone growth and density; bones in the lumbar region lose 1% of their weight per month, while bones in the hip lose 1.5% (McCarthy,

2005). It could take as much as 180 days on Earth for one's bones to return to normal. It is thought that this happens because of impaired function in the cells that build new bones, known as osteoblasts (Zayzafoon et al., 2005). The absence of physical strain and pressure due to gravity is likely the root cause of this, though it may also be associated with stress, confinement, or diet (Mack & Vogt, 1971). Furthermore, the movement of fluids causes a puffy appearance in the face and shrinking in the lower body (Williams et al., 2009).

Several NASA sponsored astronauts have reported optical difficulties following long-term missions, a claim looked into and confirmed with experimentation by Mader et al (2013). A particular astronaut, who had been on two long-term missions, was observed to have developed optic disc edema and choroidal folds. This indicates that the fluid shifts that microgravity creates within a person's head can potentially lead to the compromising of the eye's structure, especially when the person in question is exposed to multiple missions (Mader et al., 2013). Furthermore, there is evidence that exposure to microgravity may result in arrhythmias (Bolea et al., 2012). In a study done with 22 healthy males who were exposed to 5 days of head down bed rest, which mimics microgravity, experiences disruptions to their autonomic nervous system which affected the typical QRST rhythm and led to unpredictable heart rate (Bolea et al., 2012).

Olabi et al. (2002) have investigated the effect that prolonged exposure to microgravity has on the human senses, attempting to find physiological explanations for the reports of astronauts experiencing odd flavours. This may, once again, be due to the shifting of internal body fluids, or perhaps the nasal congestion that occasionally occurs due to weightlessness (Oalbi et al., 2002). This second theory seems particularly likely, as astronauts most often reported a new dislike for foods that rely primarily on scent, such as coffee.

Extreme Temperatures

Due to the harsh, and thermally extreme conditions on a spacecraft, temperature regulation is extremely important. Temperatures within the ship can climb unexpectedly if there is a systems failure, or even if the workload increases and the passengers start generating more body heat (Pisacane et al., 2007). Additionally, temperature within a spacecraft largely depends on its position relative to the sun; one side of the spacecraft tends to be lit while the other is left in the dark losing heat (Gadalla, 2005). Such temperature shifts are detrimental to astronaut performance; extreme heat could lead to muscle spasms, fatigue, or disorientation while extreme cold could lead to shivering, lack of coordination, or unconsciousness (Pisacane et al., 2007). Furthermore, many astronauts cite uncomfortable temperatures as a reason for sleep deprivation, which negatively impacts cognition, mood, and performance (Wu et al., 2018).

Stahn et al. (2017) report that the core body temperature of astronauts tends to increase over the course of a mission. During periods of activity on the spacecraft, core body temperature appears to elevate quicker than it would on Earth, and even appears elevated in times of rest. This is alarming, as maintaining a constant body temperature within a specific window is extremely important for bodily functions, circadian rhythms, mental performance, and general health (Stahn et al. 2017). The respiratory system also suffers when faced with intense heat, as one may begin to show symptoms of hyperpnea or hypocapnia (Beker et al., 2018). Hyperpnea involves increased frequency and depth of breathing (Beker et al., 2018), while hypocapnia involves the decrease of internal carbon dioxide levels (Sharma & Hashmi, 2018). Additionally, above a certain threshold, roughly 40 °C, the brain starts sustaining permanent damage as neurons are deformed and blood-brain permeability increases (Beker et al., 2018).

Astronauts are required to deal with extreme cold while performing extravehicular activities (Koscheyev et al., 2006). Though spacesuits are designed to keep them as comfortable and safe as possible, it is common to hear reports of astronauts experiencing extreme cold in their extremities. This can have serious consequences, from tissue loss, to necrosis, to possible need for amputation (Imray et al., 2009). When faced with cold, one begins to experience vasoconstriction, which, despite acting as a heat saving technique, also decreases dexterity, something extremely important on a spacecraft (Beker et al., 2018). Breathing may begin to slow down, and the brain receives less oxygen, as well as less blood flow.

Radiation

On Earth, humans are protected from the most intense radiation by the atmosphere, something not afforded to astronauts on spacecrafts.

Effects can range from acute and immediate to delayed (Hellweg & Baumstark, 2007). Acute effects may start with nausea and weakness, and escalate to a dampened immune system or anemia. Depending on the strength of the radiation, immediate symptoms can be as severe as haemorrhaging, infection, coma, or death (Hellweg & Baumstark, 2007). During low orbit missions, this is considered less of a concern, but it becomes increasingly relevant as organizations such as NASA plan more ambitious ventures (Elgart et al., 2018).

Overexposure to radiation can have devastating, and life long effects. Many astronauts, both male and female, have been found to have reproductive issues following missions, likely due to the effect high charge and energy particles have on reproductive cells such as ovarian follicles and spermatogenic cells (Mishra & Luderer, 2019). However, it appears as if ovarian cells are more sensitive to radiation than spermatogonial cells (Mishra & Luderer, 2019); it may therefore be prudent to more closely examine protective measures for female astronauts. Furthermore, another long term effect is the development of cataracts. Cucinotta et al. (2001) found a possible causal relationship between radiation and cataracts, which occurs due to genetic disruptions to eye lens cells. Cataracts have been established to be a possible side effect of radiation exposure, as noted in cancer patients and people living in high radiation zones, and appears to develop more frequently in astronauts subject to greater amounts of radiation (Cucinotta et al., 2001).

It is important to note that statistically, astronauts have a lower risk of contracting cancer than the general population, perhaps due to the additional measures space travel agencies have taken over the years to keep their astronauts healthy (Reynolds & Day, 2010). Supporting this theory, the risk of cancer related death among astronauts has definitively decreased since the 1990's. However, even if there are no immediately apparent effects due to radiation, it's important to remember that "the latent period for malignancy is 10–20 years" (Todd et al., 1999). The correlation between space travel and cancer risk is still not fully understood, largely because of small sample sizes (Hellweg & Baumstark-Khan, 2007).

Prevention and Recovery

While astronauts exercise as much as they can in-flight, to prevent the onset of crippling physiological effects, this in and of itself is not enough to make sure they can continue healthy lives once their missions are done. As researched by Petersen et al. (2017), even an astronaut who exercised 2 hours a day while in space showed clear signs of physical weakness in the days immediately after returning to Earth. While most capabilities returned to normal within 21 days, tasks involving strenuous muscle movement, such as jumping, took much longer. Similarly, diet management while in space is paramount to preventing bone loss due to microgravity; ensuring high levels of calcium and vitamin D is extremely important (Heer et al., 1999). But even this is not enough to completely solve the problem.

For impairments such as optical problems, the most effective solution seems to be implementing artificial gravity while in space (Pal et al., 2021). Furthermore, NASA has discussed monitoring the eye health of their astronauts using a goggle-like device in conjunction with skin sensors (Ansari et al., 2000). Such a system would allow for problems to be detected early, and treatment to be implemented as soon as possible. In terms of decreased bone density, recovery hinges on diet, physical activity, and medicine; for example, it appears that dried plums are useful for both preventing and repairing bone damage (Pal et al., 2021).

The European Space Agency features a rehabilitation program with a strong emphasis on physical activity (Petersen et al., 2017). Regimens are specifically designed for each individual astronaut, using a combination of physiotherapy and advice from a sport scientist. One key detail in this regimen is making sure that in the time immediately after returning to Earth, exercise happens in a pool to mimic the weightlessness of microgravity (Petersen et al., 2017). Pal et al. (2021) also recommends the inclusion of yoga or Pranayama, a technique meant for building up stamina and lung capacity, in preflight training. Though it has yet to be properly investigated, China has similarly attempted to implement Tai Chi into its astronauts routines, as a means of centering them and keeping them physically fit (Wu et al., 2018).

Finally, in terms of protection from extreme temperatures and the havoc they wreck on the human body, there is a constant push to design more effective thermoregulatory suits. Koscheyev et al. (2006) describe a suit that is able to manipulate the temperature of any given body part using compartments of water. This design also utilizes the idea of transferring thermal energy from a body part that naturally generates heat, such as the head to one that is more susceptible to cold, such as the fingers. In a later overview, Koscheyev et al. (2007)

emphasized the importance of tailoring each suit to individual astronauts, in accordance with their thermal profile.

Conclusion

In conclusion, prolonged space travel can have harsh, unexpected, and long lasting effects on the human body. The phenomenon known as microgravity can lead to many irregularities, from well documented ones such as loss of bone density, to lesser understood ones such as problems with seeing or tasting. Such impairments can take a long amount of time to recover from, if recovery is to happen at all. Furthermore, extreme temperatures and sudden thermal shifts are possibilities on every space mission, and can negatively impact performance or even endanger an astronaut's life. Extreme heat can lead to brain damage or respiratory irregularities while extreme cold is dangerous in terms of extremities and blood flow to the brain. Cosmic radiation is also a prevalent concern, with astronauts lacking the natural protection of Earth's atmosphere. This can have any number of detrimental effects, from physical discomfort, to optical disorders, to reproductive problems. It is critical to understand each and every one of these dangers, in order to better rehabilitate astronauts who have returned to Earth, or better yet, prevent the onset of such physiological disturbances in the first place.

Chapter 3

Mental Health Challenges in Space

Amna Zia

Introduction

Exploration of earth's outer space — a concept coined as space exploration or space flight — often involves long duration expeditions. Astronauts aboard space expeditionary missions have to navigate many complex and often unique stressors, for instance, malfunctions in the equipment of the spacecraft, collisions of the spacecraft with space debris, physiological disruptions in the body due to microgravity, confinement in a small area for prolonged periods of time and isolation from their loved ones (Kanas, 2016). It is crucial to understand how stressors in space impact the human psyche since compromised behavioural health has resulted in early termination of space expeditions, diminished productivity, interpersonal tension and conflict in both the US and Russian space programs (National Aeronautics and Space Administration [NASA], 2015). Moreover, as space agencies with human flight capabilities discuss plans to enable human space flight to Mars as well as the deeper solar system, long duration manned space expeditions that last many months to even years will become more common (Kanas, 2015). As the length of space expeditions increases, the probability of behavioural or psychiatric issues developing among astronauts during flight also increases (Slack et al., 2016). This chapter will therefore explore challenges related to mental health and the human mind in space.

Stressors in space

Stressors are events, or conditions of an environment that have an impact on one is optimal functioning (Kanas, 2015). Stressors in space can be physical, psychological, interpersonal, physiological and psychophysiological in nature (Kanas, 2015).

Examples of physical stressors include periods of high acceleration (of the spacecraft), microgravity (i.e., when objects are weightless), ionizing radiation, meteoroid impacts, vibration, noise pollution, uncomfortable temperatures, reduced lighting and no air or poor air quality (Kanas, 2015). Examples of psychological stressors include isolation, confined space, life-threatening danger, long stretches of monotony versus periods of high workload, personality conflicts among crewmembers, excessive amount of free time, increased autonomy and reliance on machines and local resources for survival (Kanas, 2015). Examples of interpersonal stressors include group size, limited social contacts, unfamiliarity, group heterogeneity, differences in culture and language, governance structure and leadership roles (Kanas, 2015). Examples of physiological stressors due to microgravity include motion sickness, bone loss, muscle atrophy, shifts in bodily fluids and their resulting impact on cardiac and renal function, vestibular problems and a diminished immune response (Kanas, 2015). Lastly, examples of psychophysiological stressors include sleep loss, disruption in sleeping patterns, disruptions in circadian rhythm, impairments in the sense of time, increased sensitivity to sensory stimuli, disturbances in spatial orientation, attention lapses, confusion, memory or cognitive issues, and psychomotor issues (Kanas, 2015).

It is also important to note that since astronauts partaking in space flight possess considerable autonomy, they must have the training as well as the necessary resources to handle these stressors on their own even if they communicate frequently with mission control personnel on Earth (Kanas, 2015). This is because evacuation to Earth will not be a convenient possibility for an astronaut on an on-orbit mission (Kanas, 2015).

In light of the aforementioned stressors, behavioural medicine training for the International Space Station (ISS) — a multinational space station that is in low Earth orbit — entails three significant mental disorders that one may encounter on a long duration space expedition: delirium, adjustment disorder and asthenia (Howell, 2018; Slack et al., 2016).

Delirium

During a space expedition, an acute manifestation of delirium can not only jeopardize the astronauts themselves, but also prove to be dangerous for other crewmembers (Slack et al., 2016). Delirium refers to a serious impairment in cognitive abilities that is associated with confusion, disorientation, emotional disruption and decreased awareness of one's surroundings (Badii, 2019). It is a response associated with acute illnesses, physical injury, exposure to high levels of carbon dioxide, trauma, surgery or drugs (Cunningham & Maclullich, 2013).

In the elderly population, urinary tract infections (UTIs) often trigger delirium (Cunningham & Maclullich, 2013). This is of concern to astronauts and other space travellers (e.g., space tourists) as long duration space expeditions may put one at a greater risk of developing UTIs (Slack et al., 2016). This is because the inability to urinate voluntarily is a recurrent issue in current space flight due to privacy concerns, and also potentially due to the use of promethazine for space motion sickness (Slack et al., 2016). Moreover, the microgravity conditions of space flight may also contribute to UTIs — the occurrence of kidney stones increases in environments with altered gravity, which in turn can raise the risk of developing a UTI (Weir, 2018). Other factors in space flight that may trigger delirium include hypoxia (i.e., low levels of oxygen supply to body tissues), anoxia (i.e., no oxygen supply to body tissues), inhalation of toxic gases/smoke and head injury (Cadman, 2018).

Adjustment Disorders

Two of the characteristic features of manned space expeditions are isolation and confined living conditions for prolonged periods of time (Kanas, 2015). Such living and working conditions are generally novel experiences that require time in order for one to get adjusted to them. Therefore, it comes as no surprise that common psychiatric issues associated with manned space travel are adjustment disorders, which are typically coupled with clinical symptoms of depression or anxiety (Kanas, 2016).

Adjustment disorders refer to excessive or unhealthy behavioural responses to changes or stressful events in one's life (John Hopkins Medicine, n.d.). Though anecdotal evidence suggests that most astronauts adapt well to the living conditions of space overtime, severe homesickness due to isolation from loved ones in a new environment is an adjustment reaction for some space travellers (Kanas, 2015; Slack et al., 2016). Simple countermeasures, for instance, establishing more contact with family and friends on Earth, or even sending gifts up in a resupply spacecraft, can help alleviate this adjustment dysphoria (Kanas, 2015). However, sometimes adjustment issues may disrupt one's pathology and manifest into diagnosable psychiatric disorders (Kanas, 2015). These include severe anxiety, clinical depression, as well as symptoms of other mental illnesses, ultimately necessitating the use of pharmacological or psychotherapeutic intervention — although, as discussed later, reports of diagnosable psychiatric disorders developing during spaceflight are not all that common (Kanas, 2015).

Asthenia

Space psychologists and flight surgeons from Russia have noted a unique type of adjustment reaction — called "asthenia" — to have had an impact on most of their astronauts who undertake long duration space expeditions (Kanas, 2015). Though currently not a well-defined condition, asthenia (or asthenization) refers to severe physical fatigue or tiredness as a manifestation of "nervous or mental weakness" (Kanas, 2015; Kauffman et al., 2015). Asthenia is characterized by rapid loss in physical strength, low sensation threshold (i.e., the intensity at which a sensory stimulus can be detected), severe instabilities in mood and disruptions in sleep (Boundless Psychology, n.d.; Kanas, 2015). It is thought to be a condition of cumulative fatigue that develops gradually as a result of extreme mental or physical stress and a sustained unfavourable emotional incident or conflict (Kanas, 2015). Despite space agencies following stringent criteria to assess the psychological well-being of candidates who are to partake in long duration space flight, some degree of asthenia is reported to have developed in many Russian astronauts approximately 1-2 months after flight (Aleksandrovskiy & Novikov, 1996). In this state, astronauts have decreased productivity, impairments in their sleep, anxiety, abnormal heart palpitations or perspiration, difficulties in concentration and increased sensitivity to bright lights and loud noises (Kanas, 2015).

Asthenia is derived from neurasthenia, another ill-defined neurotic psychiatric disorder that was termed to explain the response of the American public (notably the upper classes) to dramatic shifts in societal attitudes and urbanization in the1800s (Kanas, 2015; Schuster, 2003). However, asthenia that has been reported to affect Russian astronauts is believed to be a milder form of neurasthenia (Kanas, 2015). This is because these astronauts are subjected to rigorous psychiatric screening and therefore it would be unlikely for them to develop neurasthenia in space (Kanas, 2015). Another reason why asthenia in space is considered less severe than neurasthenia is because it can be countered by simple practices adopted early in space flight, (Kanas, 2015). Examples of this include sustained contact with loved ones on Earth or increased stimulation, which would prevent the need to use actual medications or psychotherapy (Kanas, 2015).

American astronauts who have participated in long duration space expeditions have also reported symptoms of asthenia (Kanas, 2015). However, both asthenia and neurasthenia have not been officially classified as diagnosable mental disorders in the Diagnostic and Statistical Manual (DSM) by the American Psychiatric Association, despite being recognized in many other parts of the world such as China, Russia and Europe (Kanas, 2015). Conditions comparable to asthenia that are recognized in America include adjustment disorders, dysthymia (persistent depression disorder), major depressive disorders, or chronic fatigue syndrome (Halverson, 2021; Kanas, 2015). Due to differences in the classification systems used to diagnose psychiatric and psychological disorders around the world, the diagnoses of asthenia (or neurasthenia) remains a point of contention internationally (Sandoval et al., 2012). Russian and American space medicine researchers are yet to reach a consensus on whether or not astronauts develop asthenia in space, and if it can be prevented or treated during long duration space flight (Sandoval et al., 2012).

Diagnosable Psychiatric Disorders

As aforementioned, it is not atypical for astronauts partaking in long duration space flight to experience homesickness as well as transient depression or anxiety (Kanas, 2015). Data collected by the Lifetime Surveillance of Astronaut Health (LSAH) program at NASA shows that symptoms of anxiety and depression have appeared during space flight (Slack et al., 2016). Similarly, data collected from NASA's Space Shuttle program (1981-2011) revealed that signs and symptoms of anxiety were reported once every 1.2 years, and signs and symptoms of depression were reported once every 7.2 years during space flight (Loff, 2017; Slack et al., 2016). If data from the shuttle program is combined with that of LSAH, it may actually show greater prevalence of symptoms of anxiety and depression among astronauts than what was officially reported during the shuttle program (Slack et al., 2016).

However, diagnosable clinical depression, anxiety or psychotic conditions (e.g., bipolar disorder or schizophrenia) that require psychotherapeutic or pharmacological countermeasures have not been reported to develop during flight (Kanas, 2015). In fact, psychiatric issues associated with human space flight seldomly occur compared to the general population, and those that have emerged have not been as severe (Kanas, 2015). Even though anecdotal evidence suggests that psychological adaptation to an astronaut's environment is more challenging on longer duration space expeditions, there have been no reports of psychiatric disorders on either space shuttle missions or ISS missions (Slack et al., 2016). To put it another way, astronauts have reported higher stress levels on longer duration space expeditions, however these stress levels have not induced clinically significant psychiatric disorders that negatively impact their missions (Slack et al., 2016).

There are many potential reasons for this. Firstly, these psychiatric conditions generally have an early onset in adult life, before the typical age at which astronaut candidates are selected, which ranges between 26-46 years (average being 34 years) (Blodgett, 2020; Kanas, 2015). An exception to this is bipolar disorder, which can occur later in adult life in predisposed individuals (Kanas, 2015). However, since astronaut candidates are rigorously screened for psychiatric and medical issues before they are selected to undertake a space expedition, it is not common for these conditions to develop in space (Kanas, 2015). Moreover, psychiatric disorders such as substance abuse or withdrawal that can alter mood or lead to psychosis, are also not readily found among astronauts due to the unavailability of addictive drugs or alcohol on space missions (Kanas, 2015).

With that being said, anyone is vulnerable to a transient psychotic reaction if they experience excessive stress (e.g., news of a tragic event such as a death of a family member) (Kanas, 2015). Therefore, it is possible for incidents of psychosis to occur during a space mission (Kanas, 2015). Moreover, even if astronauts do not have a history of depression prior to flight, they are certainly not immune from developing it during flight since behavioral health issues have been associated with highly educated and high functioning populations (Slack et al., 2016). Furthermore, it should be pointed out that signs and symptoms of psychiatric disorders may be underreported during the astronaut selection process, as well as during space flight (Slack et al., 2016). This may be caused by fear of potential astronauts that such information may prevent them from further consideration, and in the case of current astronauts, it may jeopardize their flight status (Slack et al., 2016).

Conclusion

In conclusion, astronauts undertaking long duration space missions have to deal with many complex and novel stressors, such as isolation, confinement, physiological disruptions due to microgravity and unknown danger. Despite being carefully screened for psychiatric issues, astronauts are not immune to behavioural health issues. The key psychological concerns for space travellers are delirium, adjustment disorders, asthenia, along with symptoms and signs of depression and anxiety. Although reports of diagnosable psychiatric disorders during manned space flight that require formal treatment are generally rare, it must be noted that underreporting of mental health issues by potential and current astronauts is an area of concern in space medicine.

Chapter 4

History of Space Medicine

Monica Thakadu

Introduction

First of all many people were eager to learn about space, earth orbits as well as the moon and sometimes even mars. There began Space Exploration which gave birth to Space Psychology. As decades and centuries went by, Space Medicine was then brought into the picture.
Space Medicine generally looks at the delivery of health care services and maintenance of good space environments for space travellers in different space journey durations. It looks at the preventative measures of any health outcomes from space journeys. Space Medicine has helped to curb health-associated issues faced by people going to space; be it Cosmonauts or astronauts. It has also helped to try and deeply discover how people usually get used to living on earth after going to space. Studies show that bone loss and blindness has been mainly related to space flight journeys. However, the main question is "How did Space Medicine begin?"

In this chapter, we are going to discuss the history of Space Medicine in more detail from the first mission in 1960s to Present day.

Origins of Space Medicine

Nothing yet was discovered in the early 1960s about the effects of spaceflight to human life. In May 1961 the United State president of that time, President John F. Kennedy assigned the United State to send a man to at least go to the moon and come back to earth safely. Many scientists and researchers did not know yet any health issues related to space journeys. Scientists as well, as how a human being can survive in space were asking many questions in relation to Carbon dioxide or carbon monoxide concentration and the effects of gravity in relation to motion and no one was able to get adequate answers.

That is when the Soviet pilot and a Cosmonaut Yuri Alekseyevich Gagarin got into space using a Vostok rocket, the most powerful rocket at that time. He was selected by the Soviet Union to be "The first human into space" on April 12, 1961 under Vostok 1 space mission and he was 27 years old at that time; [History of medicine in space environment/ JAXA human spaceflight].

In May 5, 1961 Alan B. Shephard became the first National Aeronautics and Space Administration, NASA astronaut to fly to space [MCGill Journal of Medicine].

According to the National Research Council, [1988] Neil Armstrong and Buzz Aldrin also took their first step into the moon. Neil Armstrong became the "first man to walk to the moon" whilst Buzz Aldrin became the second on July 20, 1969. This was the Apollo 11 crew and National Aeronautics and Space Administration hosted the journey.

Though Space Exploration started around 1960s; Space Medicine was brought into the picture during the Apollo 13 Crew. It was noticed that many astronauts after returning from space have different health related conditions such as Asthma, Deterioration of muscles/Joints and also of the brain; [Space Medicine- ESA].

NASA, the National Aeronautics and Space Administration as well as the Russian Space Agency [RSA] worked together in order to make their goal a reality. The Russian Space Programme was put into place and the International Space Station [ISS] as well as the Russian MIR Space Station were built. Their main aim was to be able to launch satellites, carry out scientific experiments on Biomedical and Microgravity issues.

As I said earlier, Space medicine was introduced after the Apollo 13 Crew. No one from the Apollo 11 Crew of 1969 has ever suffered any side effects during or after going to the moon. Except that Buzz Aldrin just suffered symptoms of altitude sickness after decades has long passed since he went to the moon; [nbc news- 12 men to walk to the moon].

It was then noticed that astronauts tend to face some critical health problems [Psychologically; mentally and emotionally or physically] depending on the duration of their space journey ; whether it was short or long. For example, it was discovered that astronauts that takes long space journeys may experience muscle wasting, bone loss, decrease in Red Blood Cell mass, altering of White Blood Cells and increased excretion of urinary calcium. The main side effects includes renal stones due to increased excretion of urinary calcium which can cause kidney failure; asthma, anaemia due to the decrease in Red Blood Cell mass, muscle and brain deterioration, insomnia, epilepsy, vomiting etc. [Human biology and Space medicine].

Headache is usually caused by dehydration whilst in space and the Space Adaptation Syndrome or radiation causes raised carbon monoxide, nausea, cough is caused by low humidity and infections.

In order to curb this health problems the following procedures were carried out:

i] Before the Journey

Well fit astronauts and cosmonauts are or were selected using the rigorous medical selection and retention process. This includes medical check- ups of lung diseases, neck problems and other chronic conditions. Age was also used as a major factor-influencing astronaut and cosmonaut selection. Psychological screening and testing were also carried out; [Science Direct].

ii] During the Journey

A] Shuttle Orbiter Medical System [SOMS]: This is a type of medical system that was carried out by the Crew Medical Officers [CMOs] and surgeons to control missions especially in shuttle orbiting journeys of space travellers; [Health: History of space medicine is centuries old/ lifestyle/daily progress].

B] Environmental Health System [EHS]: This type of a medical system focuses on the rapid monitor of radiation, air and water quality in space. Because it focuses on the air, water and radiation quality, we can shortly say that it also looks at contamination of microbes. Just like the name suggests it looks at the health wise of the environment, whether the environment is well fit enough for space travellers to be able to safely take their journeys back and forth; [Health: History of space medicine is centuries old/ lifestyle/daily progress].

C] Counter Measure System [CMS]: This one typically looks at different preventative measures regarding physiological changes that astronauts encounter; [Health: History of space medicine is centuries old/lifestyle/daily progress].

D] Health Maintenance System [HMS]: This one focuses on ways of maintaining good health of astronauts; i.e. the ways of preventing astronauts from catching diseases whilst in space and be capable to treat any if possible; [Health:History of space medicine is centuries old/lifestyle/daily progress].

E] Further advancement of space medicine: More advanced technologies in space medicine continued to be taken into consideration. NASA came up with the Digital Imaging Breast Biopsy System that was developed from the Hubble Space Telescope Technology. This system looked deeper into the woman's uterus/womb to monitor the foetus whilst in space. Remember that the Hubble Space Telescope was developed in Edwin Hubble's honor by NASA. Other developments includes the Infrared Thermometer, Memory foam, Laser Angioplasty, Ventricular Assist Device, Rocker Bagie and Space Blanket as a way of helping astronauts to cope with environmental changes. Bright lights and dark glasses were also developed and used to initiate shifting; [What breakthroughs in medicine came from NASA? / How Stuff Works].

From the above statements you can clearly see that the government, National Aeronautics and Space Administration [NASA] and the Russian Space Agency [RSA] really worked hard to instill Space Medicine into Space Psychology and Space Exploration for the well being of Astronauts, Cosmonauts or any other space traveller for their journey to be safe, sound and a success.

I believe that even today the same procedures are still practiced to ensure safety even though it seems like the National Aeronautics and Space Administration [NASA] as well as the United Nations Government has long stopped providing space journeys to astronauts and cosmonauts. Their strategy of sending people to space more especially to the moon has long stopped with the Apollo 17 crew with Eugene A. Cernan being the last man to stand on the moon in December 1972; [Gene Cernan].

Conclusion

Space Medicine has indeed provided safe and sound space journeys to space travellers in a wide range of ways that you can ever think of. It has provided a more advanced manner to save many lives. It is a "PREVENTION IS BETTER THAN CURE STRATEGY" if you wish to say so or mention that.

It all started with president John F. Kennedy making a "wake up call" for the United States government to send someone to space; [Research Gate] and a Soviet Pilot and Cosmonaut Yuri Alekseyevich Gagarin answering the call by being the "First man in history to go to space" in April 12 1961 at the age of 27 years; [History of medicine in space Environment/ JAXA human spaceflight].

The National Aeronautics and Space Administration [NASA] then found it as a form of motivation to explore the contributions of Psychology to the great space race and orbiter as well as future space journeys by sending astronauts, researchers and scientists to space on Apollo missions from sending Alan Shephard in 1961 , Apollo 11 crew of Neil Armstrong and Buzz Aldrin [1969] to Apollo 17 crew of Eugene A. Cernan and Harrison Jack Schmitt [1972].

More harm than good to space travellers was recently realised in the Apollo 13 crew and the need to explore and introduce space medicine.

NASA as well as the Russian Space Agency [RSA] join hands to make this a success with the help of the International Space Station [ISS] and the Russian MIR Space Station being built to help launch satellites and to conduct scientific research on biomedical and microgravity sciences and the Russian Space Program being implemented; [Space Medicine in Project Mercury-NASA History Division].

Regular checks before and during the space journeys were and are still done to ensure safety. This includes Psychological screening an testing as well as health systems like the Shuttle Orbiter Medical System [SOMS], Environmental Health System [EHS], Counter Measure System [CMS] and the Health Maintenance System [HMS] ; [Health: History of space medicine is centuries old/lifestyle/daily progress] as well as providing protective clothing to space travellers such as space blanket and the rocker bagie; [What breakthroughs came from NASA?/ How stuff works].

All these has helped to prevent or reduce the risks to health. These includes the risks to Coronary artery disease, renal stones, epilepsy, asthma, nausea, insomnia, deterioration of muscles and the brain, cough, headache, joint pains and alterations of White Blood Cells and reduction of the Red blood Cell mass; [Science Direct].

Chapter 5

Pathophysiology in Space

Botho Modultwe

This chapter will discuss matters that affect both the internal and external systems of the human body in space. A clear description of what the term pathophysiology means will be provided, there will also be an addition of how pathophysiology implies different body systems such as the cardiovascular system, respiratory system, musculoskeletal system and other related parts of the body. The major issue is to understand how the space environment affects the immune system.

The term pathophysiology is a combination of two words being pathology and the term physiology, which is the biological discipline that describes processes or mechanisms operating within an organism. Hence, pathophysiology seeks to explain the functional changes that are occurring within an individual due to a disease or pathologic state, (Wikipedia, 2021). According to several researches, the concept of pathophysiology seems to be changing with time with the emergence of diseases that affect humans. A typical example of the Covid-19 virus which has got scientists raising research on this concept. Pathophysiology is defined as the physiology of abnormal states specifically: the functional changes that accompany a particular syndrome or disease, (Merriam-Webster, 1828). The space environment is very extreme for the human body, long exposure may lead to health complications. High levels of radiation and microgravity may affect the human system leading to complicated health defects that may not yet have treatment for, however some defects have been proven to be reversible naturally or by scientists. It has been identified among cosmonauts and astronauts after their time in space that they have developed some changes in their bodies such as change in the structure of the skull, they can withstand some high levels of radiation than a normal human being who has never been in space and many other changes on the internal organs.

Cardiovascular System

The cardiovascular system is also known as the blood-vascular or the circulatory system. This is a system of organs that transports blood through blood vessels to and from all parts of the body carrying nutrients and oxygen to tissues and removing carbon dioxide and other related wastes, (Britannica, n.d). This system is able to manage and carry out its activities with the help of organs such as the heart with two circuits being the pulmonary and the systematic pathways. Gravity plays a major and very important role in the cardiovascular system; you can now wonder how much impact is caused in the system when one leaves earth to space. This is because as we all know that space has what is called the zero gravity where the astronauts and cosmonauts are weightless and basically nothing pulls them to the ground like for example here on earth. In an upright posture, gravity determines a pattern of fluid distribution with higher pressure in the feet and lower pressure in the head. However, in space the gradient is lost, the distribution of blood to the head causes altered responses of baroreceptor, nervous and endocrine systems. This change within a few minutes makes astronauts suffer a syndrome known as the "space motion sickness," which includes anorexia, vomiting, nausea, headache and malaise foe over 48-72 hours, (Demontis, G.C. 2017).

Even though the cardiovascular deconditioning is not extreme during space flight, it becomes hazardous on the astronauts when they come back to earth when there is a restoration of gravity. This is because the body requires immediate demand of physical capacity which cannot be met. On a research on the waveform alterations of pressure and flow rate signals in terms of the normalized signal difference, the results have shown changes that zero gravity configuration created heterogeneous stresses on the hemodynamic variables of the cardiovascular system. Volume reduction and baroreflex responses acted on global levels of pressure, volume, and flow rate, while blood shift, resistance and compliance variations altered the local mean levels of the hemodynamic variables throughout the body. However the most important alteration in volume was the combination blood shift and reduction, (Nature Partner Journals. 2020). It is not yet specifically known if a prolonged space flight mission has reversal impact on space masters when they return to space, such that they do not have to experience the same defects when they get to space again instead their body can just adapt again as it is a familiar environment. Yet the amount of liquid shift in the body caused by zero gravity when astronauts get to space has been determined in researches. About 2 liters of fluid shifts from the legs to the upper body, a reduction of around 11% reduction from the muscles.

Inside our bones, in the bone marrow we have what we call the monocytes and macrophages that circulates through the blood. Monocytes circulate in the blood as a part of inflammatory immune response. During and after its migration, both monocyte and macrophages produce many pro-inflammatory cytokines such as tumor necrosis factor alpha to further promote the immune response and recruit more inflammatory cells. However microgravity affects the microphage cells because of their unique functionality regarding inflammation and stresses within the body or their reliance on blood circulation, which can be altered in the body by microgravity conditions. It has been discovered that microgravity induces changes in macrophage metabolism, signal transduction, proliferation, cytokine secretion, differentiation, cytoskeletal structure, gross morphology, locomotion gene expression and inflammatory response, (Nature Partner Journals. 2021).

Respiratory System

The role of the respiratory system is gas exchange in the human body, this process involve the movement of oxygen, in (inhale) and carbon dioxide, out (exhale) of the lungs. The exchange of oxygen and carbon dioxide happens throughout respiratory passages involving organs such as the nose, mouth, pharynx and larynx into the lungs and the trachea, bronchioles out of the lungs, (Williams, 2011). The distribution of air in the body is fully dependent on gravity hence weightlessness has an effect on how the respiratory system function. In the experiments undertaken to test this process it has been discovered that during weightlessness there is a greater perfusion of single-breath washouts when made from cardiogenic oscillations. This was compared to an earth experiment where single-breath N2 washouts suggested a marked reduction in the inequality of ventilation distribution seen at 1G. Changes appeared on chest radiographs and regional blood flow distribution measurements. However weightlessness decreases abdominal girth, increases abdominal compliance, increases abdominal contribution to tidal volume during resting and breathing, (Engel, 1985). Looking at pathophysiology on the respiratory system, it is vital that we look at how gravity affects the major organ of this system, which is the lung. As an extremely sensitive organ, in space the lung is not affected by microgravity hence the is no any disruption on gas exchange. The one thing that may have effects on the liver and put it risk is the inhaled aerosols, (Prisk, 2005). It has not yet been discovered as to how much damage can be made on the liver due to exposure to microgravity or exposure to microgravity or radiation. But if it happens they cause change in the lungs, astronauts could have serious breathing problems in the future in their missions to Mars, (Pultarova, 2018).

Musculoskeletal System

The musculoskeletal system is a system made of organs that enables movement, support and maintain stability of an organism. This system is made up of the body's bones, muscles, cartilage, tendons, ligaments, joints and other connective tissue that support and bind tissues and organs together, (LibreTexts, 2020). The other know purpose of the musculoskeletal system is to protect the internal organs. The development of muscle and bone occurs on earth under 1G environment hence under 0G the muscles and bone goes through the process called demineralization. Under this process, both the bone and muscle lose their volume, strength and mass, which may lead to pathologies within the bone and muscle. Astronauts lose up to about 10-20% of their bone mass in microgravity, most specifically in the pelvis, resulting from altered osteocyte function within the bone matrix, (Research Gate, 2011). On other terms, the issue of bone loss leads to acceleration of age-related changes, which could be reversible on earth.in the process of demineralization, staying long enough in space may lead to development of kidney stones, bone fractures that may never be able to heal. Astronauts also face risks of loss of endurance leading to fatigue and injuries due to microgravity impact on the muscles, (Scientific American, 2005). Loss of muscles during space missions is known as muscle atrophy, which accumulates the loss of muscle strength, power and altered muscle physiology. Among the research undertaken by the ISS to address this issue, it has been discovered that crew members lose 13% of muscle in the calf muscle and a 32% decrease in peak muscle in a period of 6 months. These findings continue to change over time which led to the discovery that the long term space missions somehow helps the muscles such as the 'slow twitch' and the 'fast twitch' muscles which are responsible for maintenance of posture and endurance to adapt its self to the space environment, (Journal of Cosmology, 2010).

Other systems

Studies made have shown that space has effects on other parts of the body such as the eyes. As per NASA (2020), astronaut Barratt and some fellow crew members started having troubles reading the prescribed procedures in space, so much that they had to put on glasses. They found a symptom known as Space-Associated Neuro-Ocular Syndrome (SANS), it included swelling in the optic disc, where the optic nerve enters the retina, and flattening of the eye. Now we can wonder what really causes this eye problem, which takes us back to the shift of liquid flow to the head. The lack of gravity that increases pressure in the brain leads to changes in the eye structure hence the problem of vision. The space sight issue seems to be a still ongoing investigation, which involves both humans and animals. A typical example of the 2020 Rodent Research 23 (RR-23) mission project where 20 animals were taken on the ISS for five weeks and taken back to earth to make examinations on the various structures of the animal model's eyes. However these effects may take a period of up to five days in humans before they start reversing and going back to normal due to adaptation of vessels to the environment, (Texas A&M University, 2021).

The other system we are bound to remember is the central nervous system, how it is affected by the travel to space. The central nervous system is a network of interconnected signals that work together to transport information around the body. The central nervous system consists of the brain and the spinal cord, which are helped by neurons that help with the transportation of all the signals. Just like any other system, the central nervous system is equally affected by exposure to radiation and altered gravity in space. Exposure to microgravity affects spatial orientation, sensorimotor coordination.

Chapter 6

What are the implications of space medicine on Earth?

Kirithika Bharatselvam

You may think that medication created specifically to withstand the foreign environments of space is limited to the use of space travel. However, in reality, medication created for space missions has been able to help humans on earth in multiple ways. This is because issues that appear in space closely mimic those that humans experience on earth. For example, astronauts experience ageing changes to their physiology that mirror ageing on earth. Research that has been put into space medication also usually requires non-invasive procedures and devices to aid past space issues. This can thus be translated into medication on earth, furthering our understanding of the human body and improving terrestrial medication overall (Ruyters & Stang, 2016).

Health Research

Space medicine research has already added to our knowledge of the human body and research conducted for earth and space medication has allowed for new discoveries such as the linkage between "salt balance, nutrition, blood-pressure regulation, and bone metabolism" (Ruyters & Stang, 2016). Additionally, a multitude of analogue studies, research conducted on harsh earth environments for space missions, have been conducted. ICE environments (isolated and confined environments) (Salamon et al., 2018) used in studies for space, can also be used to further health research of those areas for the continuation of earth exploration and for earth-based medicine (Ruyters & Stang, 2016).

Preservation of the health and fitness of an astronaut is highly important in space medicine and thus, space medicine has created newer ways to mitigate the degradation of an astronaut's physical wellbeing (Ruyters & Stang, 2016). These methods can be used in earthly medicine towards rehabilitating those who are diseased or the increasingly ageing population of Canada (Ries, 2010; Ruyters & Stang, 2016). In addition, space medication has created medical diagnosis devices that are non-invasive and have been integrated into earth's healthcare system (Ruyters & Stang, 2016).

Biotelemetry

Other implications of space medication on earth include biotelemetry. Biotelemetry is a device that records physiological measurements and transmits them. This has been used during key parts of a space mission including the launch, landing of the spacecraft, and during space exploration to monitor the astronauts heart rate, respiratory patterns, and electrocardiograms onto earth (Garshnek, 1989). Biotelemetry on earth is used to treat and evaluate patients from secluded areas that may lack a variety of hospital services patients may need. For instance, biotelemetry has been used in the project STARPAHC (Space Technology Applied to Rural Papago Health Care), which provides healthcare to those located in the remote villages of the Papago Indian Reservation in Arizona, United States (Garshnek, 1989). Additionally, they use a van with healthcare equipment called the Mobile Health Clinic, to send the patient data, which may include x-rays, to physicians in hospitals where they may also be able to communicate, diagnose and prescribe treatments to patients remotely via the Mobile Health Unit (Garshnek, 1989).

Space Medicine Used to Change our Current Healthcare System

Medicine used on earth and space medicine have different overall approaches and goals. Earth-based medication primarily targets to help those who are sick, to cure or help reduce the prognosis of their illness (Ruyters & Stang, 2016). But space medicine aims to take care of patients that are already healthy, as astronauts primarily utilize medication to keep fit during the duration of their space mission. However, it is expected that there will be a shift in the earth's health care system that is similar to the concerns space medication has already aimed to treat; healthcare provided from a distance, treatment catered towards individual patients and for already healthy individuals (Ruyters & Stang, 2016). Utilizing the research done in space medication will ultimately be able to play a key role in changing the healthcare system of today's world.

The unique antigravity environment space harbours, allows researchers to expand their biomedical knowledge as they observe a human body under gravity-absent circumstances. This research can further be implicated in medical practices on earth as it can ultimately create a better understanding of the human body (Garshnek, 1989) to then create better approaches to terrestrial treatments, medication, medical technologies, etc.

The current and past research on healthcare has primarily been used to analyze and treat highly specific aspects of the human body, rather than considering the human body altogether (Ruyters & Stang, 2016). Space medicine already requires the approach of analyzing a human as a whole and catering to the specific person being treated. Thus, as space medication approaches the overall care of the human body, this research has major implications and will be increasingly useful to forming the healthcare of the future on earth (Ruyters & Stang, 2016).

Ageing

An astronaut's physiology goes through tremendous changes due to the microgravity experienced in space. These changes are occurring quickly and are mostly reversible, though, they closely mirror ageing experienced on earth (Garshnek, 1989; Ruyters & Stang, 2016). Right when astronauts experience microgravity, their immune system weakens, the lack of gravity causes risk to the neurovestibular system, the cardiovascular system becomes stressed as bodily fluids shift, bones and muscles degenerate over months or even weeks, and movement starts to become difficult (Ruyters & Stang, 2016).

Additionally, difficulties with body equilibrium control due to ageing are similar to the physiological changes astronauts face during space flight (Garshnek, 1989). Research has shown that changes in the vestibular system, central nervous system, joint receptors, and the muscles, is a result of some ageing issues. Similarly, during space missions, astronauts experience muscle atrophy, vestibular dysfunction and post-flight postural disequilibrium frequently (Garshnek, 1989).

Space studies detailing the components involved in neurovestibular changes can be used to further the understanding of the components causing similar issues on earth that are related to ageing (Garshnek, 1989). Furthermore, there have been ways to subside or slow neurological changes during space flight. These countermeasures can potentially be implemented into aiding elder patients with certain neurosensory issues (Garshnek, 1989).

Therefore, the immense research already done for space medicine can be closely applied to aiding ageing in our active and capitalistic society, which has otherwise been insufficiently understood thus far.

Pandemics

Research and medical technology from space medicine has been created to withstand and adapt to the difficulties that one may face in a spacecraft. The approaches space healthcare has already integrated into their system can be possibly applied to helping pandemics and epidemics on earth (Cinelli & Russomano, 2021). The COVID-19 pandemic involved many public health regulations being implemented including acts of isolation from others whether it be social distancing or self-isolating and self-monitorization of one's condition during a quarantine period to prevent spreading COVID-19 (Canada, 2020; Cinelli & Russomano, 2021). Likewise, being in confined and isolated environments is a condition astronauts on the International Space Station (ISS) experience regularly (Cinelli & Russomano, 2021). In addition, future space missions like the mission to Mars plan to have missions that are independent of the crew and have highly organized safety protocols and crisis management (Cinelli & Russomano, 2021). These similarities potentially imply that space medical technologies, countermeasures, and strategies can be used in terrestrial environments to limit the spread of diseases like COVID-19 and other possible pandemics and epidemics (Cinelli & Russomano, 2021).

For example, isolation and lockdown measures can become more effective when implementing 3D printing which has been used to produce pharmaceuticals and medical products (Cinelli & Russomano, 2021). The usage of 3D printing can be used to make an array of resources that people may need during pandemics/epidemics which would decrease the need for human contact (lessen the need to buy from stores). Space sleep countermeasures can also be used on earth to cope with pandemics. There are countermeasures put in place to improve sleep patterns, that can be used on earth to reduce unnecessary stresses that may be added to the pre-existing stresses caused by a pandemic/epidemic (Cinelli & Russomano, 2021).

Prevention and Treatment of Pathologies

Medicine, technology, and equipment that have been traditionally used to improve the health of individuals going to space have been proven fruitful in terms of treating a diverse range of pathologies on Earth (Orlov et al., 2014). This mainly involves the prevention and treatment of neurological pathologies as well as being a method of rehabilitation therapy (Orlov et al., 2014).

A uniquely developed suit for axial loading, known as a Regent Suit, helps improve the uptake of energy within one's body tissues as well as the movement of information coming towards the brain and spinal cord (Orlov et al., 2014). The suit itself was designed in Russia and was mainly being used for providing therapeutic rehabilitation for astronauts, after their flights in space (Cesarelli et al., 2015). The implications of this unique suit have been extremely beneficial for improving patient care while specifically improving the state of those with motor disorders which involve "focal changes in the brain" (Orlov et al., 2014). A randomized controlled trial assessing patients with history of subacute stroke but were capable of walking found that using the Regent Suit did indeed improve the "locomotion and daily living activities" (Monticone et al., 2013b). More specifically, their results found that, over the course of the study, the patients' gait speed within 6-minute walking tests increased by about 0.30m/s by going from 0.63m/s to 0.93m/s, but in the control group, it went from 0.64m/s to 0.71m/s, having only about a 0.07m/s difference (Monticone et al., 2013a, p. 796). The uses of the suit varies from improving speech and emotional status to motor function, and is being used within 28 different medical facilities worldwide to treat "post-stroke patients or patients with craniocerebral trauma, Parkinsonism, etc" (Orlov et al., 2014).

Another device that was developed based on anti-gravity suits for astronauts and cosmonauts alike for redistributing blood within the human body also has its own uses for improving patient care on Earth (Orlov et al., 2014). Its main use involves improving the prognosis of arterial hypertension as it is able to increase blood volume in the veins of the lower limbs specifically (Orlov et al., 2014). As the device can be used to lower arterial pressure, decrease cardiac output, and improve circulation overall, the device proves to be extremely valuable in improving pathologies of the cardiovascular system (Orlov et al., 2014).

Conclusion

All in all, space medicine's medical technologies, devices, countermeasures, methods, and studies have all contributed to enriching the healthcare system and research on the human body on earth. Space medicine already has a significant impact on earth-based medicine including the uses of biotelemetry for serving healthcare to secluded areas in the world and the reagent suit which improves the state of patients with motor disorders. However, the translation of space medicine into medicine on earth is practically limitless as the pre-existing research and developments of space medicine is vast and should be further implemented into earth's healthcare system in order to give the best care to humans on earth and combat issues that have yet to be fully addressed and research like ageing, and the possible medical implications that can arise from combining space and earth medicine.

Chapter 7

Medical Specialties in Space

Brianna Bedran

Presently, no surgical procedure has been performed on a human during spaceflight. However, successful tests involving anaesthesia, interventions, and survival were performed in rodents for the first time on the STS-90 Neurolab Shuttle mission demonstrating that minor surgical procedures may be a possibility for humans (Drudi et al., 2013). A variety of studies have demonstrated the feasibility of conducting extensive surgical procedures in the simulated microgravity environment and in spaceflight, including open peritoneal drainage, leg dissection, ureteral stenting, thoracotomy, thoracoscopy, laparotomy, laparoscopy, craniotomy, and microsurgery (Drudi et al., 2013). Studies have shown that manual suturing in microgravity is similar but slower than norm gravity, peak forces seem to be increased, decreased with number of tasks performed, or may remain unchanged. Subjective evaluation demonstrates that surgical procedural performance decreases when the operator is not acclimated to the microgravity environment, which may further lead to task erosion and subsequent tissue injury (Drudi et al., 2013). These weight and volume restrictions on spacecraft pose many limitations on the availability of surgical and anesthetic equipment to cover all but the most likely situations. Moreover, physiologic changes and deconditioning effects of prolonged weightlessness impact surgical diseases and treatment in predictable and unknown ways. A major surgical event will also greatly impact the mission and require a large amount of resources to be treated successfully (Campbell, 2002). Of most importance, even with the proper resources the surgical capability of any medical care system ultimately will be limited by the surgical capability and training of the crew medical officer. Due to limits on crew size and capabilities, it is difficult to have crew medical officers with the necessary intensive training to handle major surgical procedures. Still, given the increased accessibility to space and further plans for manned missions, more and more studies have been conducted on the possibility of better surgical and healthcare in space.

Trauma Management

The physiological changes that occur in the spaceflight environment will impact disease presentation, diagnostic evaluation, treatment and management. The most valuable diagnostic tool for trauma patients in spaceflight is ultrasonography using the extended Focused Assessment with Sonography for Trauma (eFAST) to evaluate the need for emergent surgical intervention. Ultrasonography on Earth is determined by gravity to locate free fluid in locations in the thorax and abdomen that can be easily detected, meaning that pathological presentations may be altered in spaceflight. Despite this, studies have demonstrated that ultrasound evaluation of several conditions can be performed successfully in microgravity, including pneumothorax and sinus fluid levels. Ultrasound-guided percutaneous aspiration of intra-peritoneal fluid using appropriate restraints was successfully performed in swine in microgravity (Drudi et al., 2013).

Advanced Trauma Life Support (ATLS) is a popular algorithmic approach to trauma management publicized by the American College of Surgeons to be used by physicians for the treatment of injured patients, with the intent to apply a simple and standardized protocol to trauma patients (Drudi et al., 2013) Recently, studies have assessed the need as well as the ability to provide special emergency care techniques to stabilize spaceflight crewmembers (patients and surgeons), restraints for fluids and equipment along with the feasibility of trauma care and minor surgical procedures. Currently, the International Space Station (ISS) has an Advanced Life Support Pack able to deliver Advanced Cardiac Life Support (ACLS) and Advanced Trauma Life Support (ATLS); however, current definitive medical management requires medical evacuation from the ISS with a delay of at least 6-24 hours if not longer (Drudi et al., 2013). Studies demonstrate that many of the common patient stabilization procedures such as including percutaneous dilatational tracheostomy, artificial ventilation, intravenous infusion using standing tubing and pressure bag system, chest tube insertion and draining, and foley insertion can be achieved in spaceflight. Drudi et al suggest that these procedures are no more difficult to perform than in the normal gravity environment, however other studies have theorized some possible obstacles with microgravity, thus the possibilities are not very clear. Still, taken together these studies demonstrate that many concepts in ATLS can be achieved successfully in a microgravity environment provided that the appropriate restraint system and equipment is available (Drudi et al., 2013).

Trauma Management

The physiological changes that occur in the spaceflight environment will impact disease presentation, diagnostic evaluation, treatment and management. The most valuable diagnostic tool for trauma patients in spaceflight is ultrasonography using the extended Focused Assessment with Sonography for Trauma (eFAST) to evaluate the need for emergent surgical intervention. Ultrasonography on Earth is determined by gravity to locate free fluid in locations in the thorax and abdomen that can be easily detected, meaning that pathological presentations may be altered in spaceflight. Despite this, studies have demonstrated that ultrasound evaluation of several conditions can be performed successfully in microgravity, including pneumothorax and sinus fluid levels. Ultrasound-guided percutaneous aspiration of intra-peritoneal fluid using appropriate restraints was successfully performed in swine in microgravity (Drudi et al., 2013).

Advanced Trauma Life Support (ATLS) is a popular algorithmic approach to trauma management publicized by the American College of Surgeons to be used by physicians for the treatment of injured patients, with the intent to apply a simple and standardized protocol to trauma patients (Drudi et al., 2013) Recently, studies have assessed the need as well as the ability to provide special emergency care techniques to stabilize spaceflight crewmembers (patients and surgeons), restraints for fluids and equipment along with the feasibility of trauma care and minor surgical procedures. Currently, the International Space Station (ISS) has an Advanced Life Support Pack able to deliver Advanced Cardiac Life Support (ACLS) and Advanced Trauma Life Support (ATLS); however, current definitive medical management requires medical evacuation from the ISS with a delay of at least 6-24 hours if not longer (Drudi et al., 2013). Studies demonstrate that many of the common patient stabilization procedures such as including percutaneous dilatational tracheostomy, artificial ventilation, intravenous infusion using standing tubing and pressure bag system, chest tube insertion and draining, and foley insertion can be achieved in spaceflight. Drudi et al suggest that these procedures are no more difficult to perform than in the normal gravity environment, however other studies have theorized some possible obstacles with microgravity, thus the possibilities are not very clear. Still, taken together these studies demonstrate that many concepts in ATLS can be achieved successfully in a microgravity environment provided that the appropriate restraint system and equipment is available (Drudi et al., 2013)

Necessity of Medical Care in Space

Extended mission durations and more and more endeavours for human exploration in space increases the necessity to deliver proper emergent medical care. There is an endless list of medical events that can be managed with surgical interventions during spaceflight including blunt and penetrating traumas (from impact with debris, during extravehicular activities, constructions and repairs, vehicle docking and refuelling, and servicing payloads), chemical contamination and burns (with electrical equipment repair, chemical and biological research, orthopaedic injuries in the setting of muscular and bone loss), minor injuries, dental complaints and other surgical pathologies that may present in spaceflight (Drudi et al., 2013).

In particularly long-term deep manned space flights, the main goal is effective functioning of a crew enclosed in a confined environment, and subjected to continuous operational and environmental stress. As humans are exposed to the conditions of space, a number of neurologic disorders can emerge. The pathological impacts during spaceflight changes a variety of neural systems ranging from motor to sensory functions, and the effects can be long lasting (Vazquez, 1998) During extended deep space flights, astronauts will be exposed to a complex environment composed of multiple acute and chronic stressors such as microgravity neurological reactions; chronic low-dose exposure of ionizing radiation; and weightlessness. Not to mention isolation, confinement, and sensory deprivation will characterize this environment, and is expected to put a heavy demand on astronauts' physiology and psychology (Vasquez, 1998).

Radiation

Space radiation is one of the foremost environmental hazards associated with interplanetary space flight. The major sources of radiation are solar disturbances and galactic cosmic rays (GCR). The components of this radiation are energetic charged particles, protons and fully ionized nuclei of all elements. Of particular concern are the high-Z and -energy (HZE) particles, with broad energy spectra at low fluence rates. Vasquez describes that HZE particles, especially Fe and its secondary fragmentation products, are of particular concern due to their high charge and energy deposition. Important ionizing radiation (IR) sources in the ISS orbits include the three primary radiation sources GCRs, which range widely from protons to Fe-ions, solar particle events (SPEs), and electrons and protons trapped in the Van Allen Belts (TPs)) outside the spacecraft. These combine to produce a complex radiation environment in and around the ISS, and the complexity of this radiation is dependent on the solar cycle, altitude, and shielding of each module of the ISS (Furukawa et al., 2020) The crews of Apollo 11, 12, and 13 reported seeing flashes of light which were later attributed to the penetration of spacecraft by high energy, high charge HZE particles producing visual sensations after interaction with the retina (Mills et al., 1983).

Radiation exposure induces many harmful biological effects, with primary concern being damage to DNA. There are various types of radiation-induced DNA damage, including base damage, single-strand breaks (SSBs), and double-strand breaks (DSBs). Among them, DNA DSBs are the most severe DNA lesions. Therefore, organisms have various DNA damage repair pathways to ensure genome stability. However, if a large amount of damage occurs or the damage is not repaired correctly, cell death, cellular senescence, and tumorigenesis may be induced (Furukawa et al., 2020).

Microgravity Reaction

Short-term exposure to microgravity produces several neurologic changes in which the Space Adaptation Syndrome (SAS) is by far, the most studied. This syndrome is reported by two-thirds of space travelers, and it is characterized by symptoms ranging from headache and stomach awareness to nausea and vomiting, beginning shortly after entry into orbit (Fujii and Pat-ten, 1992). Typically, symptoms lessen within 1 to 14 days into the flight, but the rate of recovery and specific symptoms vary widely between individuals. However, it is still unknown if the absence of symptoms reflects a complete adjustment to microgravity. It is possible that astronaut's ability to perform sophisticated tasks remains impaired long after the acute space sickness recedes. Moreover, other sensory-motor alterations, cognitive deficits, vegetative disorders, bone decalcification, muscular atrophy as well as changes in sleep-wake regulation occur during long-term space flight affecting human performance (Vasquez, 1998).

Weightless has its main damaging effects on the vestibular, musculoskeletal, and cardiovascular systems. The vestibular system is not too damaged compared to other systems despite its encounters with space sickness, producing effects similar to motion sickness. However, in the musculoskeletal system, weightlessness reduces the muscular effort required to perform physical tasks and control posture. This then produces a harmful and unavoidable demineralization of bone (Mills et al., 1983). As for the cardiovascular system, weightlessness neutralizes the hydrostatic pressure gradients in the circulation so that in particular venous pressures become uniform throughout the body. Astronauts have noted sensations of fullness in the head and nasal stuffiness during missions. Additionally, crewmen had full, distended jugular and forehead veins. This venous distension and symptoms remain throughout missions (Mills et al., 1983).

References

NASA. (n.d.). The space Medicine EXPLORATION medical Condition list - NASA technical reports SERVER (NTRS). NASA. https://ntrs.nasa.gov/citations/20110008645.

MITARAI, G. (2013, August 29). Space tourism and space medicine. The Journal of Space Technology and Science. https://www.jstage.jst.go.jp/article/jsts/9/1/9_1_13/_article/-char/ja/.

Pass, J., & Astrosociology Research InstituteSearch for more papers by this author. (2016, January 2). Medical astrosociology and SPACE Medicine: Bringing together the two branches of science. Medical Astrosociology and Space Medicine: Bringing Together the Two Branches of Science | AIAA SciTech Forum. https://arc.aiaa.org/doi/abs/10.2514/6.2016-1897.

(n.d.). Principles of clinical medicine for space flight. Google Books. https://books.google.ca/books?hl=en&lr=&id=QYrH9P0iPb8C&oi=fnd&pg=PR3&dq=scholarly%2Barticles%2Bfor%2Bspace%2Bmedicine&ots=OMkp7yrMH6&sig=bZ2LeruKXA-BiKlLm0cA5HGESZw#v=onepage&q&f=false.

NASA. (n.d.). The space Medicine EXPLORATION medical Condition list - NASA technical reports SERVER (NTRS). NASA. https://ntrs.nasa.gov/citations/20110008645.
American Psychological Association. (n.d.). Apa psycnet. American Psychological Association. https://psycnet.apa.org/record/1984-21279-001.

Lipinski, C., & Hopkins, A. (2004, December 15). Navigating chemical space for biology and medicine. Nature News. https://www.nature.com/articles/nature03193/boxes/bx2.

NASA. NASA health and medical policy for human space exploration. 2011a. [December 4, 2013]. (NPD 8900.5B). http://nodis3.gsfc.NASA.gov/npg_img/N_PD_8900_005B_/N_PD_8900_005B__main.pdf

Weight Loss in Humans in Space
Akiko Matsumoto et al., Aviation, Space, and Environmental Medicine, 2011

Campbell, M. R. (2002). A review of surgical care in space1 1no competing interests declared. Journal of the American College of Surgeons, 194(6), 802–812. https://doi.org/10.1016/s1072-7515(02)01145-6

Drudi, L., Ball, C. G., Kirkpatrick, A. W., Saary, J., & Marlene Grenon, S. (2012). Surgery in space: Where are we at now? Acta Astronautica, 79, 61–66. https://doi.org/10.1016/j.actaastro.2012.04.014

Furukawa, S., Nagamatsu, A., Nenoi, M., Fujimori, A., Kakinuma, S., Katsube, T., Wang, B., Tsuruoka, C., Shirai, T., Nakamura, A. J., Sakaue-Sawano, A., Miyawaki, A., Harada, H., Kobayashi, M., Kobayashi, J., Kunieda, T., Funayama, T., Suzuki, M., Miyamoto, T., Takahashi, A. (2020). Space radiation biology for "living in space." BioMed Research International, 2020, 1–25. https://doi.org/10.1155/2020/4703286

Lawley, J. S., Petersen, L. G., Howden, E. J., Sarma, S., Cornwell, W. K., Zhang, R., Whitworth, L. A., Williams, M. A., & Levine, B. D. (2017). Effect of gravity and microgravity on intracranial pressure. The Journal of Physiology, 595(6), 2115–2127. https://doi.org/10.1113/jp273557

Mills, F. J., & Harding, R. M. (1983). Aviation medicine. special forms of flight. IV: Manned spacecraft. BMJ, 287(6390), 478–482. https://doi.org/10.1136/bmj.287.6390.478

Schmidt, M. A., & Goodwin, T. J. (2013). Personalized medicine in human Space Flight: Using OMICS based analyses to develop Individualized countermeasures that enhance astronaut safety and performance. Metabolomics, 9(6), 1134–1156. https://doi.org/10.1007/s11306-013-0556-3

Vazquez, M. E. (1998). Neurobiological problems in long-term deep space flights. Advances in Space Research, 22(2), 171–183. https://doi.org/10.1016/s0273-1177(98)80009-4

Ansari, R. R., Singh, B. S., Rovati, L., Docchio, F., & Sebag, J. (2000, April). Monitoring astronaut health at the nanoscale cellular level through the eye. In Third Annual International Conference on Integrated Nano/Microtechnology for Space Applications.

Beker, B. M., Cervellera, C., De Vito, A., & Musso, C. G. (2018). Human physiology in extreme heat and cold. Int. Arch. Clin. Physiol, 1, 1-8.

Bolea, J., Caiani, E. G., Pueyo, E., Laguna, P., & Almeida, R. (2012, January). Microgravity effects on ventricular response to heart rate changes. In 2012 Annual International Conference of the IEEE Engineering in Medicine and Biology Society (pp. 3424-3427). IEEE.

Dunbar, B. (2012, February 15). What is microgravity? NASA. https://www.nasa.gov/audience/forstudents/5-8/features/nasa-knows/what-is-microgravity-58.html.

Elgart, S. R., Little, M. P., Chappell, L. J., Milder, C. M., Shavers, M. R., Huff, J. L., & Patel, Z. S. (2018). Radiation exposure and mortality from cardiovascular disease and cancer in early NASA astronauts. Scientific reports, 8(1), 1-9.

Gadalla, M. A. (2005). Prediction of temperature variation in a rotating spacecraft in space environment. Applied Thermal Engineering, 25(14-15), 2379-2397.

Heer, M., Kamps, N., Biener, C., Korr, C., Boerger, A., Zittermann, A., ... & Drummer, C. (1999). Calcium metabolism in microgravity. European journal of medical research, 4(9), 357-360.

Hellweg, C. E., & Baumstark-Khan, C. (2007). Getting ready for the manned mission to Mars: the astronauts' risk from space radiation. Naturwissenschaften, 94(7), 517-526.

Imray, C., Grieve, A., Dhillon, S., & Caudwell Xtreme Everest Research Group. (2009). Cold damage to the extremities: frostbite and non-freezing cold injuries. Postgraduate medical journal, 85(1007), 481-488.

Koscheyev, V. S., Coca, A., & Leon, G. R. (2007). Overview of physiological principles to support thermal balance and comfort of astronauts in open space and on planetary surfaces. Acta Astronautica, 60(4-7), 479-487.

Koscheyev, V. S., Leon, G. R., Coca, A., & Treviño, R. C. (2006). Physiological design of a space suit cooling/warming garment and thermal control as keys to improve astronaut comfort, performance, and safety. Habitation, 11(1), 15-25.

MACK, P. B., & VOGT, F. B. (1971). Roentgenographic bone density changes in astronauts during representative Apollo space flight. American Journal of Roentgenology, 113(4), 621-633.

Mader, T. H., Gibson, C. R., Pass, A. F., Lee, A. G., Killer, H. E., Hansen, H. C., ... & Pettit, D. R. (2013). Optic disc edema in an astronaut after repeat long-duration space flight. Journal of Neuro-ophthalmology, 33(3), 249-255.

McCarthy, I. D. (2005). Fluid shifts due to microgravity and their effects on bone: a review of current knowledge. Annals of biomedical engineering, 33(1), 95-103.

Mishra, B., Luderer, U. Reproductive hazards of space travel in women and men. Nat Rev Endocrinol 15, 713–730 (2019). https://doi.org/10.1038/s41574-019-0267-6

Olabi, A. A., Lawless, H. T., Hunter, J. B., Levitsky, D. A., & Halpern, B. P. (2002). The effect of microgravity and space flight on the chemical senses. Journal of food science, 67(2), 468-478.

Pal, N., Goswami, S., Singh, R., Yadav, T., & Singh, R. P. (2021). Precautions & Possible Therapeutic Approaches of Health Hazards of Astronauts in Microgravity. The International Journal of Aerospace Psychology, 31(2), 149-161.

Petersen, N., Lambrecht, G., Scott, J., Hirsch, N., Stokes, M., & Mester, J. (2017). Postflight reconditioning for European astronauts–a case report of recovery after six months in space. Musculoskeletal Science and Practice, 27, S23-S31.

Pisacane, V. L., Kuznetz, L. H., Logan, J. S., Clark, J. B., & Wissler, E. H. (2007). Use of thermoregulatory models to enhance space shuttle and space station operations and review of human thermoregulatory control.

Reynolds, R. J., & Day, S. M. (2010). Mortality among US astronauts: 1980–2009. Aviation, space, and environmental medicine, 81(11), 1024-1027.

Sharma, S., & Hashmi, M. F. (2018). Hypocarbia.

Stahn, A. C., Werner, A., Opatz, O., Maggioni, M. A., Steinach, M., von Ahlefeld, V. W., ... & Gunga, H. C. (2017). Increased core body temperature in astronauts during long-duration space missions. Scientific reports, 7(1), 1-8.

Todd, P., Pecaut, M. J., & Fleshner, M. (1999). Combined effects of space flight factors and radiation on humans. Mutation Research/Fundamental and Molecular Mechanisms of Mutagenesis, 430(2), 211-219.

Williams, D., Kuipers, A., Mukai, C., & Thirsk, R. (2009). Acclimation during space flight: effects on human physiology. Cmaj, 180(13), 1317-1323.

Wu, B., Wang, Y., Wu, X., Liu, D., Xu, D., &p; Wang, F. (2018). On-orbit sleep problems of astronauts and countermeasures. Military Medical Research, 5(1). https://doi.org/10.1186/s40779-018-0165-6

Zayzafoon, M., Meyers, V. E., & McDonald, J. M. (2005). Microgravity: the immune response and bone. Immunological reviews, 208(1), 267-280.

Aleksandrovskiy, Y. A., & Novikov, M. A. (1996). Space biology and medicine III: Humans in
spaceflight, book 2. American Institute of Aeronautics and Astronautics. https://doi.org/10.2514/5.9781624104671.0433.0444

Badii, C. (2019, August 2). What's delirium and how does it happen? Healthline. https://www.healthline.com/health/delirium

Blodgett, R. (2020, March 2). Frequently asked questions. National Aeronautics and Space Administration. https://www.nasa.gov/feature/frequently-asked-questions-0/

Boundless Psychology. (n.d.). Introduction to Sensation. Lumen Learning. https://courses.lumenlearning.com/boundless-psychology/chapter/introduction-to-sensation/

Cadman, B. (2018, January 11). Everything you need to know about anoxia. Medical News Today. https://www.medicalnewstoday.com/articles/320585

Cunningham, C., & Maclullich, A. M. (2013). At the extreme end of the psychoneuroimmunological spectrum: delirium as a maladaptive sickness behaviour response. Brain, behavior, and immunity, 28, 1–13. https://doi.org/10.1016/j.bbi.2012.07.012

Halverson, J. L. (2021, July 19). Dysthymic disorder. Medscape. https://emedicine.medscape.com/article/290686-overview

Howell, E. International Space Station: Facts, History & Tracking. Space. https://www.space.com/16748-international-space-station.html

John Hopkins Medicine. (n.d.). Adjustment Disorders. https://www.hopkinsmedicine.org/health/conditions-and-diseases/adjustment-disorders

Kanas, N. (2015). Humans in Space: The Psychological Hurdles (2015th ed.). Springer. https://doi.org/10.1007/978-3-319-18869-0

Kanas, N. (2016, June 15). Psychiatric Issues in Space. Psychiatric Times, 33(6). https://www.psychiatrictimes.com/view/psychiatric-issues-space

Kauffman, T. L., Scott, R., Barr, J. O., & Moran, M. L. (2015). A Comprehensive Guide to Geriatric Rehabilitation (3rd ed.). Churchill Livingstone. https://www.sciencedirect.com/book/9780702045882/a-comprehensive-guide-to-geriatric-rehabilitation#book-info

Loff, S. (2017, August 3). Space Shuttle Era. National Aeronautics and Space Administration. https://www.nasa.gov/mission_pages/shuttle/flyout/index.html

National Aeronautics and Space Administration [NASA]. (2015, December). NASA Space Flight Human-System Standard: Volume 1, Revision A: Crew Health. https://www.nasa.gov/sites/default/files/atoms/files/nasa-std-3001-vol-1a-chg1.pdf

Sandoval, L., Keeton, K., Shea, C., Otto, C., Patterson, H., & Leveton, L. (2012, January). Perspectives on asthenia in astronauts and cosmonauts: Review of the international research literature. National Aeronautics and Space Administration. https://ntrs.nasa.gov/api/citations/20110023297/downloads/20110023297.pdf

Schuster, D. G. (2003). Neurasthenia and a modernizing America. JAMA, 290(17), 2327–2328. https://doi.org/10.1001/jama.290.17.2327

Slack, K. J., Williams, T. J., Schneiderman, J. S., Whitmire, A. M., Picano, J. J., Leveton, L. B., Schmidt, L. L., & Shea, C. (2016, April). Evidence Report: Risk of Adverse Cognitive or Behavioral Conditions and Psychiatric Disorders, Human Research Program. National Aeronautics and Space Administration. https://humanresearchroadmap.nasa.gov/evidence/reports/bmed.pdf

Weir, K. (2018, June). Mission to Mars. Monitor on Psychology, 49(6). http://www.apa.org/monitor/2018/06/mission-mars

Canada, P. H. A. of. (2020, September 14). Coronavirus disease (COVID-19): Prevention and risks [Education and awareness]. https://www.canada.ca/en/public-health/services/diseases/2019-novel-coronavirus-infection/prevention-risks.html

Cesarelli, M., Iuppariello, L., Romano, M., Bifulco, P., & D'Addio, G. (2015). Bioengineering activities in proprioceptive and robotic rehabilitation at Salvatore Maugeri Foundation. 2015 AEIT International Annual Conference (AEIT), 1–3. https://doi.org/10.1109/AEIT.2015.7415277

Cinelli, I., & Russomano, T. (2021). Advances in Space Medicine Applied to Pandemics on Earth. Space: Science & Technology, 2021. https://doi.org/10.34133/2021/9821480

Garshnek, V. (1989). Space medicine comes down to Earth. Space Policy, 5(4), 330–332. https://doi.org/10.1016/0265-9646(89)90053-2

Monticone, M., Ambrosini, E., Ferrante, S., & Colombo, R. (2013a). 'Regent Suit' training improves recovery of motor and daily living activities in subjects with subacute stroke: A randomized controlled trial. Clinical Rehabilitation, 27(9), 792–802. https://doi.org/10.1177/0269215513478228

Monticone, M., Ambrosini, E., Ferrante, S., & Colombo, R. (2013b). 'Regent Suit' training improves recovery of motor and daily living activities in subjects with subacute stroke: A randomized controlled trial. Clinical Rehabilitation, 27(9), 792–802. https://doi.

org/10.1177/0269215513478228

Orlov, O., Belakovskiy, M., & Kussmaul, A. (2014). Potential markets for application of space medicine achievements. Acta Astronautica, 104(1), 412–418. https://doi.org/10.1016/j.actaastro.2014.05.006

Ries, N. M. (2010). Ethics, Health Research, and Canada's Aging Population*. Canadian Journal on Aging / La Revue Canadienne Du Vieillissement, 29(4), 577–580. https://doi.org/10.1017/S0714980810000565

Ruyters, G., & Stang, K. (2016). Space medicine 2025 – A vision: Space medicine driving terrestrial medicine for the benefit of people on Earth. REACH, 1, 55–62. https://doi.org/10.1016/j.reach.2016.06.002
Salamon, N., Grimm, J. M., Horack, J. M., & Newton, E. K. (2018). Application of virtual reality for crew mental health in extended-duration space missions. Acta Astronautica, 146, 117–122. https://doi.org/10.1016/j.actaastro.2018.02.034

www.ingramcontent.com/pod-product-compliance
Lightning Source LLC
Chambersburg PA
CBHW030123170426
43198CB00009B/721